Workplace Health & Safety Management Systems
D.I.Y Guide

"Do It Yourself—never outsource again"

Practical, simple, compliant solutions in developing, implementing, maintaining, evaluating and managing your own systems—never outsource again!

Benefits of building your own management systems;

- ✓ You are in total control
- ✓ You have a better understanding of the content
- ✓ You can easily make changes
- ✓ Your employees will have more buy-in / ownership
- ✓ You don't have to fork out tens of thousands of dollars for a management system that may or may not work
- ✓ Your business systems can easily integrate

Please visit **BennsonsBusinessSolutions.com** for access to many free resources including;

- – Samples / Templates / Forms
- – Information / Safety Alerts / Forums / Advice / Newsletters
- – Links to Government and Industry Agencies
- – Statistics / Trends / Innovations / Ideas
- – Contact via the website if you would like auditing, advice or support

WORKPLACE HEALTH

and

SAFETY MANAGEMENT SYSTEMS

D.I.Y Guide

JESSICA URQUHART

BALBOA.
PRESS
A DIVISION OF HAY HOUSE

Balboa Press books may be ordered through booksellers or by contacting:

Balboa Press
A Division of Hay House
1663 Liberty Drive
Bloomington, IN 47403
www.balboapress.com.au
1-(877) 407-4847

ISBN: 978-1-4525-0311-0 (sc)
ISBN: 978-1-4525-0312-7 (e)

Printed in the United States of America

Balboa Press rev. date: 11/04/2011

CONTENTS

1

WORKPLACE HEALTH & SAFETY (WHS) OVERVIEW

Workplace Health and Safety (WHS) is adopted by workplaces all over the world to keep people healthy and safe from working conditions, plant and equipment, business processes and other people.

People can be direct or indirect to a business and are defined as employees (full-time, part-time, and casual), contractors, sub-contractors, consultants, sales people, visitors, bystanders and community members.

People are kept safe and healthy by the business adopting safe work practices, given a safe working environment, provided with adequate plant, tools and equipment and offered ongoing training and education.

By law, businesses must control risks within their business to ensure people's health and safety. This is done by identifying the risk, putting controls in place to eliminate the risk, evaluating the effectiveness of the controls and educating people about the risk.

An effective workplace health and safety management system will empower employees, eliminate injuries and illness, improve the culture of the business as a whole and ensures legislative compliance.

Safe culture is a way of life and should not stop when a person leaves work.

A WHS management system is almost always included in larger businesses that have the money to invest in the system, people to manage the system but cost tens of thousands of dollars or more. Systems that are adopted for larger businesses have intricate programs and databases that can manage their data, which attracts most of the cost.

Small to medium businesses who do not have the dedicated resources or finances to implement WHS management systems sometimes do not implement anything at all. This comes from the perception that they cannot afford the expensive systems or do not have the knowledge to build the systems themselves.

WHS management systems are available "off the shelf" but mostly do not cover your specific business risks, your industry or your workplace. "Off the shelf" solutions are also expensive and can end up costing you tens of thousands of dollars—the same as a custom designed system.

The fact of the matter is if you have injuries or fatalities at your workplace, you cannot blame the person who designed your management system. The onus is on your business, management and people. So it makes sense that you, as a business have 100% input into your management systems and you have 100% control.

D.I.Y (Do It Yourself) management systems may seem daunting and time consuming—but if you have a full understanding of the process, you can better implement, review and improve your systems.

History

Workplace incidents have significantly reduced over the years as the culture of workers improves, more equipment is engineered to be safer and employers are constantly improving their systems of work to better protect people. There is still so much room for improvement as people are still suffering injuries, illness, fatalities and other significant incidents whilst at or around a workplace.

Education is the key element in improving workplace health and safety including educating workers, educating bystanders, educating manufacturers, supervisors, management and learning from history when improving our systems.

Stepping back and looking at other industries or other businesses is important to see what works and what doesn't work for example if a particular industry is still recording workplace fatalities, we shouldn't be adopting their WHS management systems. If a particular business has completely eliminated injuries and fatalities—that is the model that needs to be learnt and adapted within your business.

Providing information about "why we must do this" rather than "you must do this, because I said so" will develop understanding of the risks associated with the workplace and improve compliance in the workforce.

Understanding the levels of competency and learning techniques within your workforce is your first step to success because no matter how good your systems are, your workforce needs to understand them.

Statistics

According to Safe Work Australia's Notified Fatalities the following fatalities occurred in Australia between July 2010 and April 2011;

- **120** total work-related fatalities reported year to date *(101 workers, 19 bystanders)*

- The top incident types causing fatalities included;
  - Hit by falling object (20)
  - Crushing (14)
  - Fall from height (14)
  - Pedestrian hit by vehicle (11)
  - Trapped in machinery (10)
  - Vehicle incident—not on a public road (10)
  - Electrocution (9)
  - Vehicle incident—public roads (9)
  - Hit by moving object other than vehicle (6)
  - Drowning (4)
  - Explosion (3)
  - Hit by moving object—unattended vehicle (2)
  - Burns (1)
  - Vehicle accident—air crash (1)
  - Other—unknown type of incident (6)

- The proportion of fatalities by age group are as follows;
  - Unknown age (2)
  - 65+ years (21)
  - 55-64 (20)
  - 45-54 (21)
  - 35-44 (18)
  - 25-34 (8)
  - 15-24 (8)
  - Under 15 (2)

- The proportion of fatalities by state / territory are as follows;
  - QLD (33)—*26 workers, 7 bystanders*
  - NSW (24)—*21 workers, 3 bystanders*
  - WA (18)—*16 workers, 2 bystanders*
  - VIC (16)—*16 workers*
  - SA (11)—*11 workers*

- TAS (11)—*6 workers, 5 bystanders*
- NT (3)—*3 workers*
- CWTH (4)—*2 workers, 2 bystanders*
- ACT (0)

- The proportion of fatalities by industry are as follows;
    - Agriculture, forestry and fishing (32)—*29 workers, 3 bystanders*
    - Construction (26)—*23 workers, 3 bystanders*
    - Manufacturing (19)—*18 workers, 1 bystander*
    - Transport and storage (13)—*9 workers, 4 bystanders*
    - Mining (6)—*6 workers*
    - Cultural and recreational services (5)—*3 workers, 2 bystanders*
    - Health and community services (4)—*1 worker, 3 bystanders*
    - Wholesale trade (3)—*3 workers*
    - Retail trade (3)—*3 workers*
    - Industry unknown (3)—*1 worker, 2 bystanders*
    - Personal and other services (2)—*1 worker, 1 bystander*
    - Electricity, gas and water supply (1)—*1 worker*
    - Accommodation, cafes and restaurants (1)—*1 worker*
    - Government administration and defence (1)—*1 worker*
    - Education (1)—*1 worker*

WHS Yesterday, Today and Beyond

- Statistics are improving since yesterday
- Today there are still too many people being killed, too many personal injuries and illness
- Today we need to hold our industry authorities accountable for informing us of events, fatalities and investigations
- Today we need to strive for zero deaths, injuries and illness
- Today we need to educate all key stakeholders
- Today we need to learn from our mistakes from yesterday
- Today we need to educate ourselves with trends, techniques

and strategies
- Today we need to plan for tomorrow
- Today, tomorrow and beyond we need to always learn and improve to keep everyone safe and free from harm

2

INTRODUCTION TO WORKPLACE HEALTH & SAFETY (WHS)

Workplace health and safety requirements are broad and very reliant on your circumstances i.e. industry, risk levels, number of workers, work location etc. There are many factors that contribute to how individual businesses manage workplace health and safety including finances, commitment, industry involvement, management skills and experience, education or research.

Your commitment to workplace health and safety is required by law, so finding out how to develop systems, manage systems and evaluate systems yourselves is not only cost effective but will be specific to your workplace, current, compliant, easier to manage and maintain.

Good workplace health and safety systems are fundamental in reducing incidents, injuries and illness rates within your business. While some businesses have felt 'lucky' in the past and have essentially had a 'free kick' it is time to be pro-active and get your systems in to place and make your emphasis **ELIMINATION** and **PREVENTION** rather than a re-active approach of getting systems into place after an event or if you have been told to do so by your governing authority.

Knowing exactly what you want out of your workplace health and safety systems from the start, rather than changing the goal posts halfway through is very important and will save you time and money in the long run.

When tendering for work, you may be asked to submit your WHS Management System, or Safety Management Plan (SMP)—do not be afraid to hand it over in fears of it being 'stolen or plagiarised'—be proud of your systems, not only will it help you to secure the contract, it might just get your more work!

When you submit your SMP it doesn't have to be the whole "kit and caboodle", it should consist of the structure of your system displayed neatly in a folder that includes;

1. A copy of your policy documents (signed and printed in colour)
2. A copy of your standard documents
3. A copy of a sample procedure (the most high risk, relevant procedure)
4. A list of all procedures within the system
5. A copy of a training module, theory assessment and practical assessment
6. A list of all modules and assessments within the system
7. Copies of your JSA template, risk assessment template and incident report template
8. A copy of your asset (plant, tools and equipment) register— only assets that will be used for the job
9. A copy of your safety register (fire equipment, first aid equipment, slings, harnesses, lockout equipment)
10. Any copies of insurances, licences, employee licences/qualifications, competencies or anything else that is relevant to the job

WHS Systems Should Be;

- **Simple** *(to implement and interpret)*
- **Available** *(to all levels of the organisation)*
- **Functional** *(is realistic and usable)*
- **Compliant** *(in line with relevant legislation)*
- **Audit Ready** *(can be audited any time)*
- **Transparent** *(no hidden agendas)*
- **Prevention Driven** *(main focus on prevention)*

WHS Systems Should Cover;

- **People** *(employees, contractors, visitors, community)*
- **Planning** *(how will you achieve goals)*
- **Processes** *(all areas of the business)*
- **Production** *(the mechanical side of the process)*
- **Prevention** *(risks identified, controlled)*
- **Plant, Tools and Equipment** *(competencies, maintenance)*

WHS Systems Should Have;

- **Policies** *(top level document)*
- **Standards** *(the reason for adopting a policy/procedure)*
- **Procedures** *(step by step guidance)*
- **Forms** *(data collection tool)*
- **Checklists** *(data collection/audit tool)*
- **Audit Tools** *(tools to evaluate the systems)*
- **Training and Assessment Tools** *(tools to train/assess competence)*

Understanding WHS

Workplace health and safety, according to Wikipedia;

"Workplace safety & health is a category of management responsibility in places of employment. To ensure the safety and health of workers, managers establish a focus on safety that can include elements such as;

- *management leadership and commitment*
- *employee engagement*
- *accountability*
- *safety programs, policies and plans*
- *safety processes, procedures and practices*
- *safety goals and objectives*
- *safety inspections for workplace hazards*
- *safety program audits*
- *safety tracking and metrics*
- *hazard identification and control*
- *safety committees to promote employee involvement*
- *safety education and training*
- *safety communications to maintain a high level of awareness on safety"*

While all bases seem to be covered and there is some element of truth in this article, there are some key points that fail to receive a mention;

Workplace health and safety is the responsibility of **EVERYONE**, not just management. It is a management responsibility to ensure that there are systems in place to manage workplace health and safety.

To ensure the safety and health of workers absolutely! But what about visitors, delivery drivers, consultants, work experience students, residents in surrounding communities or the general public? Safety and health needs to cover **EVERYONE** in the immediate work area, in close proximity to the work area and in other areas that may become affected by the work area.

While all elements in the list (programs, policies, plans, processes, procedures, practices, goals, objectives, audits, hazard identification, education, committees and communication) are great tools to manage workplace health and safety, there is not enough emphasis on elimination or prevention.

Workplace health and safety should be identified as;

"Keeping people free from harm by identifying potential risks, putting controls in place to eliminate or prevent risk and evaluating the success of the controls".

A workplace health and safety system needs to be developed for a business based on their risks, not the risks of other businesses in a similar field or industry and shall include every person in and around the business.

You can start with a small circle (identifying the actual place of business) and start expanding the circle with rings to identify people or risks that are within the vicinity of the workplace. Your circle could be established using a radius i.e. within 5 metres of the boundary the risks are . . . spanning to within 5 kilometres of the business or more.

An example of this could be a nuclear power plant; their risks would include the actual workplace, then span out around 40 kilometres. In a crisis management plan for such a business they would have their worst case scenarios and who would be affected both directly and indirectly. An indirect risk would be any farms that would be in the vicinity, as the produce could be sold far and wide leaving people worldwide affected by their workplace risks.

Compare this with a home-based business providing virtual secretarial services. The risks would be at home (computer work, filing, backups, fire, computer crashing, files lost) and would also be with the people who they are providing services for (work not completed on time, work not correct).

Every business has risks and a risk perimeter; some are just smaller and less risky than others.

Be cautious of **"Out of the Box"** pre-packaged solutions i.e. purchase all your procedures for only $20,000 and start using them tomorrow! They do not fit your business operations, people or risks. These solutions may seem quick and easy at the time, but cost you more in the long run.

If you develop your management systems yourself;

- They are easy to change and keep updated
- They are easy to implement to the business
- They are easy to audit and evaluate
- You have more security and peace of mind with your systems
- The development, implementation and evaluation periods costs less

If you don't develop your management systems yourself;

- You may get taken for a ride and given a product that is not current or relevant
- You may need the provider to keep coming back when you require updates or changes
- You may have to wait a very long time for a final product
- You still have to take the time to review, audit and ensure the currency and relevance of the systems
- You may find it hard to implement or build on your systems as you are not 100% familiar
- Your management systems may differ within the business if different people / companies develop your systems and will not be consistent

3

WHS AND OTHER BUSINESS SYSTEMS

Some businesses have dedicated personnel, or whole departments that manage Human Resources, Training and Development and Safety, Health and Environment. Unfortunately in smaller businesses these functions are taken on by administrative or management staff, or not managed at all.

While you don't necessarily need a team of professionals to manage these functions, they can be easily managed administratively if the systems are simple—as long as there is a basic system in place.

Human Resources

Human resource functions should typically include;

- Recruitment and selection
- Organisational design and development
- Performance, conduct and behaviour management
- Industrial and employee relations
- Compensation and employee benefit management
- Employee retention
- Employee record keeping
- Development of position descriptions and KPI's

- Provide advice and guidance to personnel regarding remuneration and benefits, industrial relations or equal employment opportunity
- Provide advice and guidance to management regarding industrial relations, recruitment and performance management
- Prepare and maintain the Human Resource Management Plan

Training and Development

Training and development functions should typically include;

- Induction, safety, equipment and process training, assessment and awareness
- Developing training plans, training needs and gap analysis
- Coordination of external training and assessment requirements
- Provide advice and guidance to personnel regarding training, competencies, career path and training and development initiatives
- Provide advice and guidance to management regarding personnel competencies, skills gaps and training requirements
- Prepare, maintain and evaluate the Training and Development Management Plan

Safety, Health and Environment

Safety, health and environment functions don't always, but should typically include;

- Provide advice and guidance to training department regarding safety, health and environment topics
- Provide advice and guidance to management regarding incident, injury and illness trend analysis

- Provide advice and guidance to personnel regarding safety, health and environment procedures, practices, hazards, risks and initiatives
- Conduct local risk assessments and incident investigations
- Maintain registers of safety equipment
- Conduct audits of people, processes, plant, tools or equipment to ensure compliance
- Prepare, maintain and evaluate the Safety Management Plan

What should typically be involved in human resources, training and development or safety, health and environment is not always what actually happens. In some cases either the personnel do not have the knowledge, experience or time to completely fulfil the requirements of the function, leaving some very important factors completely forgotten or unmanaged.

For example, in many organisations recruitment is the only function performed by human resources personnel. This leaves a huge gap for other critical aspects of human resources such as retention or planning.

Training and development is sometimes limited to personnel only filing paperwork and keeping a basic system and no time developing people.

Safety, health and environment is generally very understaffed (if at all) and undermanaged. Most personnel in a safety, health or environment capacity would agree that most time is spent being re-active. What this means is there is little or no time for pro-active measures including audits, trending or initiatives to eliminate or prevent incidents, injuries or illness.

In hindsight, if the other two functions (human resources and training and development) were spot on, there would be more time for safety personnel to be more pro-active.

These three systems must work together for the same purpose— healthy, safe, competent people!

4

INCORPORATING WHS
IN BUSINESS FUNCTIONS

The business structures (outlined above) all have a critical role to play in their business functions including people, process and plant, tools and equipment. This section will outline how the three areas must work together in achieving a successful management system;

- Human Resources (HR)
- Training and Development (TD)
- Safety, Health and Environment (SHE)

People

The following outlines how people are managed within the business and what areas are responsible for this function.

Pre-Employment

- Development of position description (HR, TD, SHE)
- Establishment of role KPI's (HR, TD, SHE)
- Development of training plan (HR, TD)
- Management of recruitment and selection (HR)

Employees

- Management of personnel induction, internal training and safety awareness training (TD, SHE)
- Management of personnel competencies (TD)
- Management of personnel skill development (TD, SHE)
- Management of workplace risks, hazards and incidents (SHE)

Non-Employees (Contractors, Visitors, Students)

- Management of induction, training and safety awareness (TD, SHE)
- Management of workplace risks, hazards and incidents (SHE)

Community

- Management of risks, hazard and incidents that may affect the community (SHE)
- Management of communication and education provided to the community (TD)

Process

The following outlines how business processes are managed and what areas are responsible for this function.

- Management of process requirements for specific departments, areas and job roles (HR, TD, SHE)
- Management of competency and compliance requirements (TD, SHE)
- Management of personnel competencies (TD)
- Management of risks involved in the process (SHE)

Plant, Tools and Equipment

The following outlines how plant, tools and equipment are managed what areas are responsible for this function.

- Management of plant, tools and equipment requirements for specific departments, areas and job roles (HR, TD, SHE)
- Management of competency and compliance requirements (TD, SHE)
- Management of personnel competencies (TD)
- Management of risks involved with tools, plant and equipment (SHE)

Outsourcing

Typically in a business, if there is something that needs doing and you want it done quickly you would outsource. Before deciding to outsource you need to weigh up the benefits, based on your constraints, of either outsourcing your management plans or doing it yourself.

Cost—What is the price to outsource your management system? Would it be cheaper to hire someone on a contract basis to do it cheaper? Would it be cheaper to have a current employee develop the system?

Time—How long will it take? Will the provider be able to complete it within the specified timeframe? How long would it take a current employee to do it part time?

Currency—If you outsource, will the information be current? Will the provider be giving you information that they implemented a decade ago? Will the information be compliant to relevant legislation?

Accuracy—If you outsource, does the provider know your business or risks? Would they spend any time at all on your site? Are they selling you information that has no relevance to your business? Would someone internally be a better judge of your internal risks?

Final Product—If you outsource, will your final product be professional, easy to maintain or update? Will the provider charge you every time you want changes made? Is the final product good enough to submit to an auditor without any input from you?

Free Resources

Like web development, everything available to build, maintain and support business systems is freely available. The only constraint is time—the time it takes to research requirements, search for templates and actually build your systems.

Free resources are available from your industry authority, WorkCover, Safe Work Australia and many other government agencies that were development purely to help businesses like yours. There are good and bad experiences with using freeware to develop your systems—but each individual business will have their own constraints i.e. time, money, compliance or resources.

The biggest concerns with free information are currency and accuracy—how current is it? Is it an article from 2004 that is no longer current with legislation and standards? Is the information right? Can you trust the person who is providing the information?

As a business risk, you need to assess the validity of information that you want to use, so the best location to source your information is the authority who would audit your business i.e. if you are setting up a health management system, the best place to gain guidance is the WorkCover Health Management Toolkit located on their website, this way your system is "built to comply" from the beginning.

Consultants and Specialists

Outsourcing consultants and specialists can be tricky. Not only do they cost money, they may not have the skills or experience that they claim and may end up costing your business more money than anticipated.

As a business, you need to put a consultant through the paces like a pre-employment interview. You are going to be paying them a lot of money to provide a service, so you need to ensure that all your business needs are met. Ask the provider to present their strategies to your management so that questions can be asked and samples, timelines, quality and content can be discussed.

Find out what work they have done in the past, speak with existing customers (if available), ask for samples and ask for a service agreement contract to protect your investment.

The Service Agreement / Contract Should Include;

- ✓ Timeframe
- ✓ Total Cost including development, implementation and support
- ✓ Compliance guarantee
- ✓ Qualifications of persons who will be working on your systems
- ✓ Copy of their current Professional Indemnity Insurance
- ✓ Payment charter (try and include at least 50% payment at the start and 50% upon successful completion of job)

People

Your duty of care is to provide a safe work environment for all people associated with the workplace including direct involvement and indirect involvement, whoever falls within your risk perimeter.

Risks can be prevented and eliminated by;

- Identifying risks
- Controlling risks
- Evaluating the effectiveness of controls
- Conducting regular risk assessment
- Identifying and managing change within the business

People at the Workplace

People that are directly associated with your workplace include;

- Employees
- Contractors
- Delivery drivers
- Visitors
- Consultants
- Sales reps

Risks associated with people in the workplace include;

Injuries—injuries from equipment or process

Illness—illness from working conditions or external conditions

Trauma—trauma from personal incidents or incidents of others

Risks associated with people directly involved in the workplace are the responsibility of management to identify, control and evaluate effectiveness of controls. Workplace induction, training and awareness and review can prevent and eliminate most risks in the workplace.

People Near or Around the Workplace

People that are indirectly associated with your workplace include;

- Pedestrians passing by
- People in vehicles passing by

- People in close proximity of the workplace
- Community members in areas surrounding the workplace

Risks associated with people near or around the workplace include;

Injuries—injuries from equipment or process

Illness—illness from external conditions

Trauma—trauma from business incidents

Risks associated with people indirectly associated with the workplace are harder to manage. It is the responsibility of management to identify, control and evaluate these risks to ensure the safety of the whole community.

A good example may be a business approaching a school to talk to the children if a fatality has affected one of their peers or expanding your Employee Assistance Program (EAP) and offering it to community members if required.

Skills and Competencies

A person's skills, experience and competencies need to be managed throughout their entire employment, even before a person is employed to when they leave.

- ✓ A position description will identify what is required for a role including skills, knowledge, experience, safety, process and production KPI's.
- ✓ A training plan will identify what training and education is specifically required for a role in the order that is required.
- ✓ Ongoing scheduled training (internal and external) will be required for initial training and refresher training.

✓ Ongoing health and safety awareness is required including consultation, communication and initiatives provided to people.

Human Resources—will identify skills, experience and competencies required for the role, recruit and appoint someone who has the minimum requirements and provide personnel with a career path within the business.

Training and Development—will induct personnel with specific risks, collect copies of certificates, qualifications and licences and offer training and assessment on process, plant, tools or equipment.

Safety, Health and Environment—will ensure people are aware of risks, hazards and procedures that exist, monitor and review behaviour to ensure processes are working, consult with personnel when developing new systems.

Safe Behaviour

Regardless of a person's skills, experience or competency levels, there is always an element commonly referred to as 'human error', but is centred on personal behaviour.

Safe behaviour in a business is collectively known as the safe culture. Safe culture and individual safe behaviour must start from the top of the business and flow through to the workforce and other people at the workplace. The goal for any business is to create a safe place of work by enforcing individual safe behaviour, which will create an overall safe culture, which will result in a safe business.

Safe behaviour is instilled though communication, education, training, hands-on workshops and on the job mentoring. People need to be included in the process to gain ownership and fully understand the why's of safe behaviour.

Safe culture is instilled through the same methods; however the emphasis should be on people looking out for one another and everyone being involved in improvement strategies.

A safe business should acknowledge and celebrate safe behaviour at any chance they get. People respond to positive reinforcement and acknowledgement.

Process

Each process within a business has inherent risks to people, production, plant and equipment. To successfully get the best output from people, training and education is required. People should understand the risks that exist, the purpose of the control and the relationship of each of the processes within the business. A better understanding will create a more ownership and compliance.

Process risks not only include what can hurt people, but also include damage to equipment, cause production loss or environmental risks.

It is the responsibility of management to;

- ✓ Identify every risk that exists in each process
- ✓ Consult with employees (or persons directly involved in the process)
- ✓ Implement controls to eliminate or prevent the risk
- ✓ Communicate risks and controls to entire workforce
- ✓ Evaluate the success of the control
- ✓ Provide instruction (policies, standards, safe work procedures, training)
- ✓ Evaluate instructions (theory assessment, practical assessment, task observation)
- ✓ Identify new or changed risks when changes to the process occur
- ✓ Constantly review the process

Risk Assessment

A risk assessment will identify risks within the process again, not only risks that hurt people but risks that will impede production. A risk assessment can be performed the same way for any task within the business and a simple template should be used so that people are familiar with the standard risk assessment form and technique that should be used.

An important point to remember is potential or hidden risks must be identified, not just the visible risks. This is called 'crisis management' where all potential risks are identified within the whole business.

Basically a full risk assessment should include the following steps;

1. List every **step** required to perform the task
2. Identify the visible or potential **risks**
3. Identify how a risk will be **controlled**
4. Identify how a risk would be **managed**
5. **Plan** the job based on the risk assessment
6. **Evaluate** the effectiveness of the risk assessment—did it work?

It is imperative that the process actually reflects the risk assessment. There is no use in a complete and thorough risk assessment sitting in someone's drawer and not available while on the job.

Evaluating the risk assessment, after the job or process is complete is imperative to better improve the risk assessment process i.e. where new risks identified during the process that weren't identified during the risk assessment?

Controls

Once a risk has been identified, a control must be put into place to protect the health and safety of all people. The hierarchy of controls is a simple methodology outlining the process that should be followed when controlling risks starting from the top (harder controls) and working your way through to the bottom (softer controls);

Elimination

Substitution

Isolation

Engineering

Administration

Personal Protective Equipment (PPE)

In terms of risks, your best chance for permanent processes is to eliminate the risk completely. If you only 'reduce' the risk, it still exists and you still have that obvious chance that something will go wrong.

If a risk cannot be eliminated completely, that is when you work your way down the "hierarchy of controls".

For example; could you use a different tool? Can you cordon off the area completely to bystanders? Could you complete the job after hours? Can workers were different PPE? Can we use more people or resources?

In the following scenario, a worker has to complete a simple task where obvious and hidden risks exist.

Task:

The worker is installing signage outside the office block. A drill and a ladder are required for the task.

Risks identified:

- Water present on ground—*electrocution from drill cord hanging in water*
- Uneven ground—*injury from falling off ladder*
- Signs are heavy and awkward to lift into place—*signs or drill could be dropped onto worker or passers by*
- Electricity is present in wall—*electrocution when drilling into wall*

Elimination—Can the risk be completely removed?
YES—The worker identifies that a cordless drill should be used to eliminate the risk of electrocution.

Substitution—Can the process be done differently to eliminate the risk?
YES—The worker identifies that the signs could be placed lower, to eliminate the need to work from heights.

YES—The worker identifies that the task requires two people—1 person to install the signs and 1 person to pass the signs.

Isolation—Can the source of the risk be removed?
YES—The worker identifies that the power source to the building should be isolated for the duration of the task.

Engineering—Can the process be redesigned to eliminate the risk?

Administration—Can policies, standards or procedures be implemented to eliminate the risk?
YES—Add all the controls into a standard safe work procedure for future tasks.

Personal Protective Equipment—Can people use added PPE to eliminate the risk?

YES—People that are walking around the area must be wearing hard hats.

Safe Work Procedure

If the task or process is repetitive it makes sense to develop a standard safe work procedure, so that in the future people do not have to carry out a risk assessment for the task, they just follow the safe work procedure.

When developing safe work procedures, it is essential to include people that are part of the process to ensure that all steps are included. It is also a good idea to use the risk assessment and physically step-out the process when writing the procedure.

A safe work procedure should have the following headings to ensure all items are covered;

- **Purpose** (why has the procedure been developed?)
- **Scope** (what area does this procedure cover, who shall comply?)
- **Referenced documents** (what documents were referenced in this procedure?)
- **Definitions**
- **Prerequisite Competencies** (what does a person need to already have?)
- **Tools, Equipment and Materials Required**
- **Personal Protective Equipment (PPE) Required**
- **Procedural Details** (task broken down step by step)
- **Appendices, acknowledgement** and **sign-off**

The best way to develop a work procedure is to write it in a way that you could teach a child; this will remove any complacency within the procedure e.g. you may know what a 'stack' is and where to find it, but does a new person know this information? Be clear and concise with all information.

The safe work procedure should have a unique identifiable document number and should be available so that it can be found easily by people who require a copy.

All procedures and other business documentation should be reviewed at least every two years to maintain currency and accuracy.

These documents can be managed within the business with a Document Management System.

Training and Assessment

Using safe work procedures, you can develop assessment tools such as;

- **Training modules** *(a series of procedures that directly relate to the task)*
- **PowerPoint presentations** *(a visual display/demonstration of the task)*
- **Training videos** *(a visual display/demonstration of the task)*
- **Theory assessments** *(asses a person's knowledge and understanding of the task)*
- **Self-assessment** *(person can assess their own skills and ability)*
- **Practical assessment** *(assess a person's skill and ability to carry out the task)*

Trainers—Anyone can train other people, as long as they are currently competent in the task that they are training.

Assessors—Ideally an assessor is someone who has a current Certificate IV in Training and Assessment. In the case that a person does not have that qualification they also must be currently competent in the task that they are assessing.

It is ideal to use a different trainer and assessor for workplace training.

The theory assessment should ask key questions that are important to the person's health and safety. There is no point asking questions like "What time is lunch"? This is evidence that serves no purpose to your management system or the person's health and safety. Some good questions to ask are;

- Where is the emergency evacuation point?
- What is the emergency telephone number?
- Where are the first aid supplies located?
- What does D.R.S.A.B.C.D stand for?
- What actions would you take if you heard the evacuation siren?

Ask questions that show a person's knowledge in safety, health and environment for your workplace.

The theory assessment can be completed while the person reads a document (module), watching a PowerPoint presentation, watching a video or listening to an instructor. Be sure to include information in the materials so the person can answer the question.

Once the assessment is completed, have a discussion with the person and assess the questionnaire in front of them—this way if they have answered any questions wrong you can talk about it. They may have only got one question wrong, but that one question could be imperative to the location of something or important instructions. If that person commits a breach or has an injury it is good to go back to the questionnaire and confirm that they did know the answer—it won't look good if you are being audited and their answer was wrong and not corrected or assessed.

The practical assessment should cover every step in the task in the correct order and should be highly focused on technique, safety and protocol. You can add knowledge-based questions in your practical assessment and ask questions while they are performing the task.

Some good practical assessment tasks could include;

- Can perform a pre-start and can explain the purpose of a pre-start.
- Can locate the emergency stop.
- Can start up the machine as per the correct procedure.
- Can perform the task as per the procedure.
- Understands why the task must be completed in the order prescribed.
- Can shut down as per procedure.
- Understands why equipment is washed at the end of the shift.
- Can wash the machine at the end of the shift.

It is important that the steps are clear, concise, realistic and achievable. You can give the person a copy of the practical assessment while they are in training so they can better prepare for the assessment or do a self-assessment.

It is also very important not to set people up for failure for example including items in the practical assessment that are simply not achievable for a beginner, something that takes years to master. This will result in wasted time for everyone. You can have an experienced person perform this difficult task in front of the trainee, so they can gauge where they need to be in the future.

Assessments should be dated and signed by all parties i.e. participant (trainee), trainer, assessor and training or human resource personnel.

Some businesses develop comprehensive training modules which cover a range of procedures for a specific job role i.e. Truck Operator. This is a very time consuming and long process, but very rewarding when training many operators.

Some businesses also go another step further and either purchase or develop their training modules around the national competency units to give their workers nationally recognised training, again this is a long, time consuming and expensive process.

Records that must be kept for training and assessment;

- Copies of licences and qualifications
- Copies of completed assessments (theory and practical)
- Copy of training hours (evidence towards competency)
- Copies of training or assessment evidence (items used for training or assessing such as pre-starts, workplace inspections)
- A spreadsheet matrix or database of training and assessments that are completed, required or expired and due

Don't let Registered Training Organisations (RTO's) scare you into believing that your employees need "Nationally Recognised Training"—check with your industry or governing body before you spend thousands of dollars on training that you could have done in-house.

Nominal Durations

It is recommended that all processes, procedures and training have standard nominal durations that are included as pre-requisites, training hours and time as a competent operator. Having adequate training durations in place will reduce incidents and injuries and will ensure that people are not rushed through the business too quickly and they have sufficient time spent on a task to become experienced operators.

For example, a truck operator might have to be in the business doing other jobs for at least three months before training on a truck is offered. The person must then spend at least 200 hours training on the truck before they can be assessed. The person might then have to have a minimum of one year as a competent truck operator before they can advance to the next level.

While experienced people can have incidents or injuries, new people are just as at risk of hurting themselves or others due to lack of knowledge or inexperience.

You can record a person's training hours on a simple spreadsheet, a training log book or in a database. The training hours should be included as evidence of training in the person's training file.

The training phase, if possible should be done with a few different trainers so the person gets a good feel of how the task is done by everyone.

It is a bad idea to have the theory assessment assessed by the same person who provides the training and conducts the practical assessment. The problem here is if the person training has bad habits or unsafe behaviour it is inherited by the trainee. The trainee will pick the bad habits up and be absolutely unaware that they are performing the task in an unsafe manner.

This is difficult for some businesses to provide multiple resources, so if you must use the same person for the training and assessment process pick someone who has the right values and is also a good performer. As a business owner you can always audit the assessment process by sitting in occasionally and watching the person who is being trained or assessed.

What evidence should you have by the end of the assessment process?

- ✓ Copies of licences, qualifications, resume or evidence from a previous workplace
- ✓ Copy of a completed, signed-off theory assessment
- ✓ Copy of training hours completed (including trainer sign-off)

✓ Copy of a self-assessment if done
✓ Copy of a completed, signed-off practical assessment

Plant, Tools and Equipment

This section will not guide you on your specific plant, machinery, tools or equipment; however it will give you easy to understand ways in which to manage your assets. If you plant, machinery, tools and equipment are 100% serviceable, your employees know how to use it safely and efficiently and essentially improve your bottom line.

It is a great idea to keep a register of your plant on a single spreadsheet, this way you can monitor inspection, insurance or registration due dates. You can also include servicing, repairs and maintenance and leasing / finance if applicable.

If you inadvertently forget to re-register a machine or vehicle and it is involved in a collision due to miss management of paperwork it could cost the business dearly. Outlook is also a great tool to add due dates as they will remind you well in advance when payments are due.

Have you got a pre-start system in place for your vehicles or machinery? You can purchase (or make your own) basic pre-start sheets or books to list basic to critical items on vehicle that must be checked prior to operation.

Having these tools alone will not work unless training is provided. Training employees to carry out a pre-start check and advising not to operate if faults are identified, it is unsafe or to tag out for repairs. This can be done via a toolbox talk, training on the pre-start form and a practical demonstration on how to carry out a pre-start check on a vehicle or machine.

Tools are one of the most important assets of your business. Tools can include cash registers, mechanical tools, trolleys, barcode scanners and other important items that keep your business running smoothly.

It is also a good idea to keep a register on a spreadsheet which includes date of purchase, cost, supplier contact details, scanned copies of your receipts and a replacement date. The replacement date should never be "when the tool dies or becomes useless", the replacement should happen;

- When you are having a good month and can afford it
- Long before it is unserviceable
- Before you sell the tool being replaced
- When you can pick up a bargain from your supplier (whilst buying something else from them)

Like any part of the business, if you plan well in advance the impact is going to be minimal, if at all if the tool fails at a critical time.

Equipment can include so many facets of your business for example if you are a catering business; your equipment would include aprons, mobile phones, first aid kits, fire blankets, food supplies, sauces or condiments. Equipment is just as important as vehicles, plant, machinery or tools of your trade—if the equipment fails or runs out it could have serious impacts on your business, time and sales.

Again, you could keep all of this information on a spreadsheet register to ensure that you can keep a record of expiry dates, quantities, price, supplier etc.

While registers can take some time to put together, they can save you a lot of time and money in the long run because they will keep you ten steps ahead and will only take minutes to update them regularly. Your accountant will love it too.

- When you are being audited, your registers can be easily emailed to the auditor.
- When you are updating your insurance you can easily provide the insurance company with your register which will include supplier, purchase date, price, quantities,

receipt numbers, warranty information and other important information.

- When employees are enquiring about your safety equipment, you can quickly provide them with a copy of your register including equipment type (fire extinguishers, first aid kits, eye wash stations, and spill kits), expiry dates and current locations.

Production

Your product is your business; the speed in which you produce a product or service, the quality, the design or the execution. The people, process, plant, tools and equipment are what keeps production going, so good Human Resource, Training & Development and Workplace Health & Safety systems will ensure that your production doesn't suffer, you don't lose time with product damage, incidents, illness or injuries.

Regardless of the industry, business size or business risk, all management systems can be managed the same way.

Risk Assessments don't only determine the likelihood and consequence of injuries; they can also be used to determine the financial risks of lost production, down time or damaged equipment.

Example;

If you identify that production will be at a standstill if your truck brakes down, the control that you would put in place would be to have another truck on standby whether it be a truck in the workshop for service, hire truck or outsourcing to a contractor. This is probably referred to as good planning (or good luck), however if the cost of the broken down truck is critical to your business, I would call this crisis management, which would be picked up in a JSA, risk assessment or crisis management plan.

Workplace Health and Safety, Human Resource or Learning and Development Management systems are not unlike other successful systems in your business such as the systems you use to schedule, budget, schedule resources, warehousing or tendering—as a matter of fact, if other systems in your business are successful you should have great success developing any new systems.

5

DOCUMENTATION

Documentation is what formalises the management system. The documentation within a management system should be structured in a simple manner which will make communicating painless and fulfil your compliance requirements.

The structure should look similar to this (in this order);

- Policy
- Standards
- Procedures
- Data collection tools (forms, checklists, registers)
- Safety related tools (JSA's, risk assessments, MSDS)
- Compliance tools (audit tools, reference materials)
- Training and assessment tools (PowerPoint presentations, training modules, technical information, videos, assessment tools)
- Evaluation tools (questionnaires, surveys, feedback forms, observation forms, audit tools)

To simplify a management system, you could break each area down and give them their own section for example a Safety, Health and Environment management system can comprise of three sections;

1. Safety
2. Health
3. Environment

This would eliminate confusion and make the systems easier to implement. Each section would then have its own set of documents that include policies, standards, procedures, forms and other materials.

Policies

A policy is a statement from the company. They are the overarching documents which have standards, procedures and forms sitting below them and should be simple statements, not procedural information.

Policies should be signed by the General Manager (or highest authority) and proudly displayed at the place of business and included on the business website, inductions or employee handbook.

This document is the **"What"** of the management system. What is required from the management, what is required from the workforce and what is required from external bodies such as visitors or contractors?

Please visit **BennsonsBusinessSolutions.com** for samples, examples and templates that you can download and use as policy documents.

Try visiting some business websites and looking at their policies as you will get a good feel of how others write their documents and what types of policies they have adapted.

Your governing authority will also be able to tell you what policies you must have and what they should include.

An example of a Policy;

Injury Management Policy

The company is committed to the health and safety of all personnel at this workplace. Injuries will be managed as per the requirements of the Injury Management and Workers Compensation Act 2004.

This will be achieved by;

- Appointing a Return to Work Coordinator
- Managing all injuries with a return to work plan
- Allowing injured personnel to select their own medical practitioner
- Assisting the injured employee to return to work within the specifications of the return to work plan
- Managing injured workers fairly and consistantly

Signature
Full Name
General Manager
Company Name

Standards

A standard is the reason why policies or procedures exist and what it was developed around i.e. Australian Standard, Code of Practice, Acts or Regulations.

This will ensure your business is using best practice as your guideline to developing policies and procedures.

This document is the **"Why"** of the management system. Why have we implemented this strategy, why must we follow these rules, why are we looking out for employee's health and safety?

If you were to tell an employee that they must do a task in a certain way and they asked why, the resistance would show if you answered "because I told you to". It is human nature to know why we must do things a particular way.

So, when you implement a new procedure or protective clothing for example, if people knew that it would better protect their health and safety from a certain chemical or risk, the resistance and compliance is less of a challenge.

A standard that accompanies a policy or procedure will give the employee more information and will reinforce why the policy or procedure has been put into place.

An example of a Standard;

Injury Management Standard

1. **Purpose**
   The purpose of the Injury Management Standard is to comply with the Workers Compensation Act 2004.

2. **Scope**
   This standard must be adhered to by all personnel.

3. **Referenced Documents**

   ▪ Injury Management and Workers Compensation Act 2004
   ▪ WorkCover Guide 123—Injury Management

4. **Standard**
   4.1 Choosing a Medical Practitioner
   Injury Management and Workers Compensation Act 2004—Section 4—*"Injured persons have the right to choose their own medical practitioner."*

Procedures

A procedure is a set of steps a person must follow to ensure the policy and is upheld to the required standards.

This document is the **"How"** of the management system. How do we perform the task, how do we know what protective equipment to wear, how do we know what risks have been identified in the task?

A procedure can include;

- Technical information
- Diagrams
- Photographs
- Flow charts
- Bullet points
- Other instructional information such as operator manuals

There is no ambiguity within the workforce when there are clear procedures that must be followed to complete a task. Every person must complete the task in the same manner and follow the same instructions when doing so.

The procedures must cover;

- Identified risks and controls
- Step-by-step instructions
- Personal protective equipment (PPE that must be worn)
- Roles and responsibilities where applicable

An example of a Procedure;

Injury Management Procedure

1. **Purpose**
   The purpose of the Injury Management procedure is to ensure

injuries are managed to the required standard

2. Scope
This procedure must be adhered to by all personnel.

3. Referenced Documents

- Injury Management and Workers Compensation Act 2004
- WorkCover Guide 123—Injury Management

4. Procedural Details
4.1 Notification of Injuries
- Injuries must be reported to the Supervisor as soon as possible.
- Injured persons are to report to the Injury Management Coordinator during the same shift that the injury occurred.

4.2 Return to Work Plan
- A Return to Work Plan will be developed for the injured worker, in consultation with the Injury Management Coordinator, the Supervisor and the injured worker.
- The Return to Work Plan is binding and must be followed.

Data Collection Tools

Data collection tools are used to gather information regarding a certain process or requirement.

These can be presented in many different fashions, it very much depends on usability, functionality, difficulty or how the information must be relayed back. Pros and cons of some data collection tools;

Printed Forms—Pros

- Simple and easy to use

- Available to anyone within the workplace
- Familiar for people who are less computer literate or do not have access to a computer

Printed Forms—Cons

- The forms are uncontrolled when printed which means that if an update or change is made to the form old versions may still be used
- Sometimes handwriting can be difficult to interpret
- People outside of the workplace have difficulty completing forms
- Not every person in the workplace has access to a computer, making it difficult to access electronic forms

Electronic Forms—Pros

- Easy to complete, consistent and easy to read
- People outside of the workplace can complete and email them

Electronic Forms—Cons

- Cannot obtain a signature unless printed
- Must have an intranet / internet to access
- Fields can be locked to particular formatting, data types, dropdown boxes – making it easier for the user to complete

Spreadsheet—Pros

- Quick and easy to use
- Can easily manipulate, filter, lookup, chart or table data
- Very powerful when comparing or calculating information

Spreadsheet—Cons

- Can become too large and store too much information
- If not set up properly can store multiple records and can become confusing
- People can easily make changes to the data

Database—Pros

- Easy to enter data using simple forms
- Can export data to many different formats, including excel
- Can lock the database so people can't make changes unless they are authorised
- Parameters can be set so records can't be duplicated

Database—Cons

- You will need someone with database experience or training to manage the database, forms, tables, reporting and data
- Difficult to maintain if it is a purchased database, you may have to pay for changes.

Safety Related Tools

Safety related tools are used more so as reference materials when developing procedures and for monitoring on-the-job safety.

- A risk assessment or JSA will be used to write a procedure or as a stand-alone document that workers use as they do a job.
- Pre-start checklists, workplace inspections and other on-the-job safety forms are not only used to help people identify safety issues, but they are sometimes also used as data collection tools for other purposes i.e. pre-starts given to maintenance so they can schedule maintenance.

- Hazard reporting forms make good safety tools as people use them to identify, control and report any hazards or risks that they encounter whilst on the job – they serve no purpose if nothing is done with the data after they are handed in.
- Incident report forms, investigation checklists and risk matrix are tools used to manage incidents, accidents, investigations and the process used to determine the root cause and contributing factors of an incident.

These tools must be developed around the management system. There is no point collecting data that you will never use and it makes no sense not collecting data that you need to fulfil the requirements of your management system.

All of your tools need to be mapped back to the policy, standard, procedure and assessment tools to ensure all bases are covered.

What you should consider when developing Safety Related Tools;

Risk Assessment

- The risk assessment tool needs to be easy to use and actions clearly defined on the form.
- There needs to be enough space on the form for people to write their steps, risks and controls.
- This form is best laid out in columns with clear bold headings, tick boxes, highlighted areas and action options.

JSA

- The JSA is a commonly used form and should also be very easy to use and understand by all levels of the business.
- There needs to be enough space to write on the form and is also best laid out in columns like the risk assessment.

- The approval area of the form should be at the top so names and signatures approving the JSA are visible and clear.

Pre-Start Checklist

- The pre-start does not need to be huge; it can be developed on a small scale and have a series of tick boxes.
- There needs to be a section at the top large enough for a name, date, shift and piece of equipment that they are going to operate.
- There needs to be a bold sentence at the bottom of the form that states **"This equipment must not be operated if faults are found – Tag out and report faults or damage to your supervisor"**.
- It is also a good idea to include on the form when, who and where it needs to be handed in to.

Workplace Inspection

- This form should include specific items within your business that are required to be checked.
- Any high risk items that have been identified for your business must be in the workplace inspection.
- The form needs to have simple tick boxes for ease of use.
- There needs to be clear instructions as to when a person may not work in the area i.e. if a 'No' is ticked work must not commence.
- An area at the top must include enough room for name, date, shift, location and area of the workplace inspection.

Hazard Report Form

- This form can be generic and the same form can be used by different businesses or industries.

- The form needs to include an area for name, date, shift, location of the hazard and details of the hazard.
- It needs to include a column for any controls that have been put into place already.
- It also needs an area for actions that need to be done at a later date or time. This area needs a column for action, responsible person and due date.
- It is important that any outstanding actions from the hazard report form be managed and information communicated back to the person who reported the hazard and others working in the area.

Incident Report Form

This form can also be generic and should include the following fields;

- Incident type (injury, damage, production loss, theft, non-compliance)
- Injury classification (first aid, medical, lost time)
- Dates (incident date, reported date)
- Name, crew and company of persons involved including the supervisor or any witnesses
- Details of the incident
- Risk ranking (likelihood and consequence based on your risk matrix)
- Actions taken during and after the incident to prevent reoccurrence
- Other actions to be completed at a later date including responsible person and due date

There are so many more fields that can be included in the incident report form. This will depend on what information you want to collect and how thorough you want this information. Your governing authority will have templates and samples of an incident report form and what is required to be collected.

Incident Investigation Checklist

Any items that will be covered during your investigation must be included on the list. This will assist when gathering information including;

- Witness statements
- Pre-start checklists
- Workplace inspections
- Photographs, drawings or diagrams of the incident or the area
- Medical treatment notes
- Other information relevant to the actual incident

Risk Matrix

The risk matrix must be suited to your business. There is no point copying another business risk matrix as your consequences might be different either in the financial or production risks. The matrix must include the likelihood of the event occurring and the consequences if the event occurred. You don't have to have a huge matrix, just as long as it suits.

Likelihood

- Rare
- May Occur
- Expected to Occur

Consequence

- Low
- Medium
- High

There needs to be information against each of the levels (likelihood and consequence) so they can be ranked correctly.

For example;

A 'Rare' occurrence in your business may be an 'Expected to Occur' in another business or a 'High' consequence may be a production loss of $1,000 for your business but may be $1,000,000 for another.

Comprehensive training must be provided to the entire workforce if these tools are going to add value to your systems.

Compliance Tools

Compliance tools are used to benchmark your tasks, processes, or systems against the standard or legislative requirements. The best audit or compliance tools that can be used to compare your systems are the tools that are used by your governing authority i.e. WorkCover this will ensure that your systems are kept to a standard that should always comply.

Internal compliance and audit tools can be simple spreadsheets, forms, databases or other platforms. Results from audits should be regularly used to improve your systems such as monthly, quarterly, annually this will greatly depend on the size of your systems.

When mapping your systems out to your governing authority's audit tool you should include a list of;

- The policies that you are expected to have in place
- What you must comply with i.e. Australian Standard, Code of Practice, Acts or regulations
- What quality are they looking for
- How often have they advised revision of your documents or systems
- What should you have in place to monitor any changes within the industry or governing bodies i.e. are you subscribed to their newsletters or do you attend meetings?

- What key words will they be looking for i.e. consultation, communication, risk, WHS
- Does your audit tool mirror theirs? This will make it easier to audit

Training and Assessment Tools

Training and assessment tools are used to measure a person's knowledge skills, and competency on a particular subject, task or operating a piece of equipment.

Tools can include;

- PowerPoint Presentations
- Videos
- Demonstrations
- Verbal Training by an Instructor
- Training Manuals
- Technical Manuals
- Theory Assessments
- Self-Assessments
- Practical Assessments

Application of the tools can include;

- Classroom One-on-One Training
- Classroom Group Training
- On-the-Job One-on-One Training
- On-the-Job Group Training
- Simulated Training
- Open Book Assessment
- Strict Closed Book Assessment

These tools need to be kept as evidence of competency and need to be accurate, accessible and need to include;

- Trainee's full name, date and signature
- Trainer's full name, date and signature
- Assessor's full name, date and signature
- Date of training or assessment
- Name of training or assessment being performed
- Assessor to tick or initial written answers or practical steps
- Competency result clear i.e. circle/tick Competent or Not Yet Competent
- Assessor to include if further training is required and explain these details clearly

Results of assessments need to be communicated back to the trainee and their supervisor. Common situations that occur are a trainee completes a theory assessment at home and drops it back to work. The assessor does not get around to marking the assessment straight away.

The person and their supervisor assume that they are right to go ahead and train on the equipment or task because they have handed in their assessment.

The assessor later identifies many mistakes in the assessment and has to call the person back for more training.

The problem – it wastes time for everyone involved, the trainee is operating a piece of equipment or completing a task before an assessment is marked and could pose a risk to themselves and others.

Evaluation Tools

Evaluation tools are used internally to measure how your systems are working. Audit tools are similar to measure compliance, but evaluation tools are to measure performance and success of the system for example a compliant system may not be used by workers as it may not have been communicated correctly or to difficult to execute.

The evaluation will gather data, monitor performance, gain feedback and compare with other systems—continuous improvement of the systems is the goal. If systems do not work or are not 100% functional or practical, the system should go through a review process and be changed to reflect the less successful factors.

These tools can be formal or informal;

- Checklists
- Forms
- Surveys
- Questionnaires
- Observations
- Comparing old data with new data to measure improvement
- One-on-one interviews with employees
- Evaluation of other business systems compared to yours or other industries practices

6

ROLES AND RESPONSIBILITIES

Everyone within the business has roles and responsibilities. Without certain roles, the management system will fail. A person can take on many of the roles and responsibilities or many people can share roles and responsibilities, as long as everything is covered. Roles are identified below in **bold** followed by their responsibilities in the development, management, compliance and evaluation of the WHS Management System.

Process Owner—this is the business or the department within the business that is responsible for the process for example SHE Department.

Manager—this is the person in charge of the department responsible for the process for example Michael Smith—Administration Manager.

Author—this is the person who wrote the document i.e. policy, standard, procedure.

Reviewer—this is the people who have been part of the review process. These people must have some involvement with the process.

Document Controller—this is the person who is responsible for ensuring the current version of documents are made available to people either via an intranet, computer system or hard copies. In many small to medium businesses this role is taken by someone in administration or reception.

Everyone—all people are to comply with systems, policies, standards and procedures. Everyone should have input (consultation) in the development and review of documents.

Contractors—contract companies and their employees or sub-contractors are responsible to follow and comply with your systems.

Trainer—this is the person who is responsible for induction and training of your people. The trainer must be a good communicator, competent in what they are training and have a good safety culture—this is important as this person is the first person that new people will have contact with.

Assessor—this is the person who is responsible for assessing competencies. In small to medium businesses this role is taken by the trainer or an employee. This person must be thorough, fair and reasonable as they will be the people who deem someone "competent" in a particular task.

It is important that certain roles are consistent throughout your management systems—this will ensure that everyone has a good understanding of who is who. If half your documents refer to the "Head of Department" and the other half of your documents refer to "Department Manager" it will show inconsistencies and a lack of trust in the systems.

While something like this may seem unimportant, people do recognise inconsistencies and will start to question the actual content of the documents. This works the same way with naming of equipment, processes and terminology—everything has to remain consistent.

When developing documents you might add specific roles and responsibilities for example;

Document: Dust Monitoring Procedure

Process Owner—WHS Department
- *Responsible for maintaining the procedure and ensuring it is current, relevant and communicated to the workforce.*

Manager—WHS Manager
- *Responsible for approving the content of the procedure.*

Author—Health Coordinator
- *Responsible for developing the procedure to ensure it complies with legislative and industry requirements.*

Reviewer—Environmental Officer, Safety Officer, WHS Committee and Employees
- *Responsible for reviewing the content and making recommendations as required.*

Note: Each document will have specific roles and responsibilities depending on the task or process. It is very important not to "copy and paste" roles and responsibilities throughout your documents—you can't hold people accountable if their role or responsibility is not clear and precise.

7

MANAGEMENT SYSTEM TOOLS

A management system is likened to a "well oiled machine" when all the tools are used correctly. It doesn't make sense to have one extremely good management system and no effort put into the others. The following tools should make up each of the management systems;

Crisis Management Plan (CMP)

A Crisis Management Plan is about the whole business—what are the worst case scenarios in every aspect of the business? Once all the scenarios are identified, you work your way backwards to eliminate, control and prevent the risks.

A CMP does not need to be carried out often, however if you have major changes within the business it is advisable to review the plan and update as required.

Your entire Workplace Health and Safety Management System will rely on this plan. All of your documents and practices within the system will be for one single cause—to eliminate and prevent every single item on the CMP, so it is important to keep this document handy and audit your own systems to ensure compliance with your CMP is always maintained.

A great way to start the CMP is to get your key stakeholders in your meeting room for the day and start talking about your worst case scenarios, start writing them down and grouping them into categories.

You might want to order lunch in for this meeting because it may be a long one. Have a whiteboard, markers and eraser handy and have someone taking notes.

You can group your big ticket items into categories and sub-categories such as;

- Natural disasters (earthquake, flood, snow, cyclone, hurricane etc.)
- Fire (chemical, hydrocarbon, explosion, machinery, office etc.)
- Fatality, injuries, illness (falls, hit by objects, vehicle impact etc.)
- Vehicles and Machinery (break downs, availability of parts, availability of repairers, theft etc.)
- Bomb threat (evacuation of workforce, communication to workforce, police involvement)
- Hostage situation (police involvement, communication / evacuation of workforce, training for reception personnel etc.)
- Terrorist attack (items of value on site, evacuation procedures, police involvement etc.)

Crisis Management planning can become quite involved and everyone can add valuable input into the planning process. Many of the main category or sub-category items will have the same or similar controls – just make sure that there is a control.

Noting is unrealistic or silly when it comes to crisis management because if the 'highly unlikely' event does occur one day you will be prepared and if it never happens that's even better!

Risk Identification and Management

Risk identification is about one area of the business, or one process. Much the same as crisis management, you work your way backwards to ensure all the risks are covered.

Once you have identified a potential threat or weakness in the business through the CMP, you can complete a risk assessment to establish the scale of the risk.

You will step out every task associated with the process, implement controls to eliminate or reduce the risk and you can use this information to develop procedures used for training.

Job Safety Analysis (JSA)

As described earlier, a JSA is a mini risk assessment and is used to identify risks of a certain task.

This should also be carried out if a risk has been identified in the CMP. This will micro-manage the task and clearly identify specific risks and controls in the task.

If the task is performed often, a procedure should be written so personnel do not have to generate a JSA every time they do the task. These can also be used to develop a procedure and training tools.

8

CONSULTATION

It is a requirement that everyone has a role to play in developing, reviewing and maintaining management systems. The consultation process should include;

- Formal and informal discussions with people (including employees and contractors)
- People being able to make suggestions for example suggestion box, open forums, feedback forms
- Discussing WHS items with people at toolbox meetings, committee meetings or training days

Why?

If your employees have input into the development of your management systems, policies, procedures or processes they will have "buy-in" and more ownership. If employees are essentially responsible for an action or process there tends to be more compliance as it hasn't been handed down from management, it has been decided by their peers.

9

COMPLIANCE

Compliance is a scary word and leaves everyone wondering if they do comply, what do you have to comply with and how do you become more compliant?

The answer is easy if you follow these steps;

1. Find out your governing authority i.e. WorkCover
2. Use their auditing tools to map your systems against what is required
3. Develop a gap analysis to outline what needs to be done
4. Set yourself a timeframe to get these items done
5. Re-audit your system using the authority's tools (as per step 2)
6. Get an independent audit done on your system
7. Have the authority come out and evaluate your system

Other things you can do is go online and see what other people are doing—but be careful as some of these systems are not correct and you may be putting your business at risk by spending time implementing a system that does not fit your business or your governing authority i.e. an American business system may not fit an Australian business.

10

COMMUNICATION

Communication is a process where, in relation to business processes, people within the business are briefed about a change within the business for example implementation of a new policy. When the communication process happens it is imperative that information is communicated;

- to everyone involved, not half of the workforce and definitely not only to management
- effectively and everyone understands the information given
- in a timely manner with sufficient time for consultation i.e. not one day before a change is implemented

As already stated, you could have the best management system in the world, you could have paid thousands of dollars for it and had it developed by the best consultants but, if you do not have a good communication system the process will have been done in vain.

People have different ways of learning for example some people learn from reading, writing, verbally or orally. It is important that all learning types are included when communicating.

Some good communication systems can include;

- Posters on noticeboards or sent out with payslips
- Notices put up on your internal intranet or website
- Verbal at toolbox meetings, shift change meetings or training days
- One-on-one with different departments, areas or crews
- Developing "Communication Days" away from the workplace

Some other good extended communication systems, for compliance reasons can include;

- Individual memos that are signed off by people and added to their personnel file
- People signing an "Attendance Sheet" at meetings or training days
- Questionnaires that are completed by people on a particular topic

If you are planning to implement something within the business that will have different effects on people, it is important to have phases for example;

Phase 1—Initial Communication (1-2 weeks)

Phase 2—Consultation (2-4 weeks)

Phase 3—Feedback Communication (1-2 weeks)

Phase 4—Implementation (2-4 weeks)

Phase 5—Enforcement

This will ensure people are given fair warning and plenty of adjustment time when changes are implemented in the business.

A good example of this is implementing a "No Smoking" policy at your workplace. You may have some unhappy people if they show up for work one day and the site has gone from a smoking allowed site to a no smoking site overnight.

A good way to implement this is to communicate the reasons why the site is being changed to no smoking, give people the opportunity to discuss and share their views and give fair notice of the implementation date i.e. two months.

You will have more compliance and less disgruntled employees if the communication and consultation processes are done correctly.

11

IMPLEMENTATION

Once the planning, communication and consultation processes are complete it is time to implement the change.

The implementation process or the "roll-out" process may take some time especially for people to adjust to the changes. Patience must be observed during the implementation phase as some people will take longer than others to settle in.

The implementation phase must have a clear timeframe so people know what date the change will be enforced. This timeframe (5 phases) should be communicated at the beginning during the communication phase.

Changes should be implemented as smoothly as possible with the least amount of disruption to people. It is in human nature to resist change however if people are forewarned, consulted in every aspect of the change, communicated with at every turn and given adequate time to adapt the changes implementation can be a very successful process.

Note: If the change is implemented smoothly within the given timeframe other changes within the business will be accepted more freely. If the change is messy and is only implemented by halves your credibility will suffer and people will be resistant to other changes.

If your business has very few systems in place it probably isn't a good idea to implement many systems at once. Not only will the workload be enormous, but you may get some resistance as people may not be used to structure and organisation.

If you are in this situation, a good approach would be to give yourself a year or two to implement multiple systems into the business. Start with a plan which includes realistic timeframes in which you can implement each of the systems.

If done correctly your systems will have longevity and will be appreciated throughout the workforce.

If you associate with other business owners find out how long it took for them to develop and implement their systems – most will tell you that they are still working on it and have a long way to go.

You may even find yourself reviewing systems every two years which will be a continuous process of implementation and review that never ends. This is good—it's progress.

12

EVALUATION

A management system is no good if it doesn't work.

How can you tell if it is working? Auditing or evaluating each element of the systems will identify what works and what doesn't.

Some great ways to self-audit your systems are;

- **Analysing** trends (incidents, injuries, equipment damage) are they rising, stable or falling?
- **Communicating** with the people who use the systems— are they happy with the systems? Do they have any other improvement ideas?
- **Benchmarking** pre-change data to post-change data—is production up or down? Have costs changed? Is there more or less time spend completing tasks?
- **Interactions / Observations** with people—watch people performing their work, talk to them and have a discussion about the process, equipment or environment.

The evaluation phase should take place sooner than later. The sooner after the implementation phase the better (preferably before the enforcement phase) this will ensure that the change is not enforced and then changed again in a short period of time.

There is no reason why a change can't skip through the phases i.e. implementation phase, then back to the communication phase and over and over again before you get to the enforcement phase.

You may as well get it right the first time, and you get to keep your integrity.

13

AUDIT READY

The way in which you developed your management system (referring to your governing authority and their audit tools, planning, communication, consultation, implementation) it should be "Audit Ready".

To guarantee that your system is compliant with the governing authority you need to have a copy of their audit tool, the document that they would refer to when auditing your system.

These should be available from their website. Some tricks to make your systems audit-friendly include;

- ✓ Align your naming convention to theirs (your documents are called the same as theirs for example Monitoring of Airborne Contaminates). This way when they go looking for the equivalent it is easily found.
- ✓ Align your elements to theirs (each heading within your documents, or elements can be called the same as theirs for example Communicate in the Workplace). Again, when the governing authority is looking for your coverage of a topic it is easily found.

- ✓ Use the governing authority's examples as your template, the auditor will be familiar with the format and layout and will be able to run through your systems very quickly.
- ✓ Map your systems to the authority's audit tool (you can use a spreadsheet to identify each of the items that will be audited then identify what documents cover each item).
- ✓ This will make self-auditing and external auditing a breeze. It will also help if legislation changes and you can easily identify what documents you will need to update.

14

SUCCESS

The success of any business should not only be based on production—the fact that there have been no injuries (or reduction), no property damage (or reduction) or a low turnover of staff should be celebrated also.

The people factors (human resources, workplace health & safety and learning and development) have a huge role to play in the success of your people, their health and wellbeing and their happiness.

There are many initiatives that can be introduced into the business to retain, inspire, improve performance and reward performance. There is no limit on the number of initiatives that you implement, as long as their effectiveness can be measured and evaluated.

Some examples of initiatives are;

- Safety incentives (LTI bonus, safe behaviour rewards, safety improvement rewards)
- Health initiatives (gym memberships, health and wellbeing campaigns)
- Rewards and recognition (attending training days, process improvements)

- Remuneration and benefits (bonuses, extra annual leave, cash in sick leave)

Managing Success

It is important to share your success with the people who made it happen including employees, customers and other stakeholders. It is human nature for people to respond to positive reinforcement. It costs nothing to acknowledge the good work, good attendance or good team work of people, as long as it doesn't make others in the team feel unimportant.

There are many ways you can tell people that their efforts don't go unnoticed or that they are not appreciated including;

- Monthly rewards and recognition (fuel cards, gift cards) for excellent performance
- Annual celebration (BBQ, dinner, event) for excellent performance
- Team building days (golf, bowls, workshops) as a reward
- Monthly newsletter identifying outstanding performance or safe behaviour
- Mentioning a few names during a Friday smoko

The best part about managing success is you can do it as cheaply or as expensively as you desire, you can make it very public or as private as required as long as the employees feel rewarded and everyone in the business is reaping the rewards of the success.

Communication is the key to managing success—let people know if you have had a good month or a bad month, give people the chance to act accordingly.

Communication can be by way of;

- Noticeboard announcements
- Emails

- Newsletters
- Website notices/announcements
- Notices sent out with people's pay slips

I am sure the families of workers would feel very proud of their loved ones if they got a personal letter from the GM thanking them for their efforts in the past month.

15

PLANNING FOR THE FUTURE

Like preparing a budget for next financial year, or planning the output of your operation over the next ten years, it is important to factor in success into your planning. If you can budget an initiative or reward into your financial planning, then communicate it to your workforce they will have all year to look forward to the rewards.

You could even establish a social committee and match them dollar for dollar on an extravagant dinner on a private yacht, holiday, end of year party or even new watches for everyone.

There are so many ideas and you don't have to get them of the internet or from a HR consultant, find out what your workforce want, then plan for it.

Your people are your best assets and basically your business wouldn't be where it is without people. Telling or showing people how much they mean to the company is what makes you the "employer of choice". Anyone can give a pay rise or shout smoko but publicly acknowledging someone will go a very long way.

Planned successes can be based around;

- Zero injuries
- Zero equipment damage
- Zero downtime
- 100% attendance to training days
- Compliance with policies, standards, procedures
- Team players, team leaders
- People caring about each other—pat on the back awards
- People's choice i.e. a person nominated by the workforce

THE SKY IS THE LIMIT

I wish you every success in the development and implementation of your management systems. Be confident in what you have achieved thus far and don't be afraid to start something new. Please do not be hesitant to get your employees involved with any phase of your management plan, you will find that they have some of the best input—they are the ones that do the work and know the risks.

If you would like an independent audit of any of your systems for compliance or piece of mind don't hesitate to contact me – my door is always open or subscribe to my mailing list at **BennsonsBusinessSolutions. com** for launch dates of other books in the series including Document Management, Health Management, Incident Management and Risk Management.

www.ingramcontent.com/pod-product-compliance
Lightning Source LLC
Chambersburg PA
CBHW022129170526
45157CB00004B/1810